About Us

Hi! My name is Malia, I'm one of two people behind T-Bot, the most curious Wondroid. My creative husband, Johnathan, created this lovable tiny robot back in 2010 when he was only 15 years old! A decade later, in 2021, we were ready to share T-Bot with the world. Our mission is to re-spark exploration of the world through the lens of wonder.

With T-Bot and the Wondroids, we strive to create opportunities where kids can learn and develop at their own pace. We believe that our activity workbooks are helpful tools for parents who want their children to learn and grow with confidence in themselves.

I'm so proud of what we have achieved thus far. I truly believe you and your kid will fall in love with T-Bot just as I did when my husband first showed the tiny bot to me!

If you enjoy our books and want to help us share our mission, please leave a review - we'd really appreciate it! Thanks for being here!

Part 1:
0-49

BONUS Section:
Coloring Break

Part 2:
50-100

Free Goodies?

Scan this QR Code

Luminis Creo LLC

Number Tracing Workbook

First Edition
ISBN: 978-1-960423-23-8

AF887977

This brilliant book belongs to:

T-Bot wants to give you this very brilliant award!

psssst printable color version :)

0 zero

How to use this workbook:

1. Follow the doted lines.
2. Trace each number.
3. Keep practicing in the blank spaces!

(This is an example of a filled out page.)

0 0 0 0 0

0 0 0 0 0

0 0 0 0 0

zero zero

zero zero

zero zero

0 zero

0 0 0 0 0

0 0 0 0 0

0

zero zero

zero zero

zero

1 one

1 1 1 1 1

1 1 1 1 1

1

one one

one one

one

2 two

2 2 2 2 2

2 2 2 2 2

2

two two

two two

two

3 three

3 3 3 3 3

3 3 3 3 3

3

three three

three three

three

4 four

4 4 4 4 4

4 4 4 4 4

4

four four

four four

four

5 five

6 six

6 6 6 6 6

6 6 6 6 6

6

six six

six six

six

7 seven

7 7 7 7 7

7 7 7 7 7

7

seven seven

seven seven

seven

8 eight

8 8 8 8 8

8 8 8 8 8

8

eight eight

eight eight

eight

9 nine

9 9 9 9 9

9 9 9 9 9

9

nine nine

nine nine

nine

10 ten

10 10 10 10

10 10 10 10

10

ten ten

ten ten

ten

11 eleven

11 11 11 11

11 11 11 11

11

eleven eleven

eleven eleven

eleven

12 twelve

12 12 12 12

12 12 12 12

12

twelve twelve

twelve twelve

twelve

13 thirteen

13 13 13 13

13 13 13 13

13

thirteen

thirteen

14 fourteen

14 14 14 14

14 14 14 14

14

fourteen

fourteen

15 fifteen

15 15 15 15

15 15 15 15

15

fifteen

fifteen

16 sixteen

16 16 16 16

16 16 16 16

16

sixteen

sixteen

17 seventeen

17 17 17 17

17 17 17 17

17

seventeen

seventeen

18 eighteen

18 18 18 18

18 18 18 18

18

eighteen

eighteen

19 nineteen

19 19 19 19

19 19 19 19

19

nineteen

nineteen

20 twenty

20 20 20 20

20 20 20 20

20

twenty

twenty

21 twenty-one

21 21 21 21

21 21 21 21

21

twenty-one

twenty-one

22 twenty-two

22 22 22 22

22 22 22 22

22

twenty-two

twenty-two

23 twenty-three

23 23 23 23

23 23 23 23

23

twenty-three

twenty-three

24 twenty-four

24 24 24 24

24 24 24 24

24

twenty-four

twenty-four

25 twenty-five

25 25 25 25

25 25 25 25

25

twenty-five

twenty-five

26 twenty-six

26 26 26 26

26 26 26 26

26

twenty-six

twenty-six

27 twenty-seven

27 27 27 27

27 27 27 27

27

twenty-seven

twenty-seven

28 twenty-eight

28 28 28 28

28 28 28 28

28

twenty-eight

twenty-eight

29 twenty-nine

29 29 29 29

29 29 29 29

29

twenty-nine

twenty-nine

30 thirty

30 30 30 30

30 30 30 30

30

thirty

thirty

31 thirty-one

31 31 31 31

31 31 31 31

31

thirty-one

thirty-one

32 thirty-two

32 32 32 32

32 32 32 32

32

thirty-two

thirty-two

33 thirty-three

33 33 33 33

33 33 33 33

33

thirty-three

thirty-three

34 thirty-four

34 34 34 34

34 34 34 34

34

thirty-four

thirty-four

35 thirty-five

35 35 35 35

35 35 35 35

35

thirty-five

thirty-five

36 thirty-six

36 36 36 36

36 36 36 36

36

thirty-six

thirty-six

37 thirty-seven

37 37 37 37

37 37 37 37

37

thirty-seven

thirty-seven

38 thirty-eight

38 38 38 38

38 38 38 38

38

thirty-eight

thirty-eight

39 thirty-nine

39 39 39 39

39 39 39 39

39

thirty-nine

thirty-nine

40 fourty

40 40 40 40

40 40 40 40

40

fourty

fourty

42 fourty-two

42 42 42 42

42 42 42 42

42

fourty-two

fourty-two

41 fourty-one

41 41 41 41

41 41 41 41

41

fourty-one

fourty-one

43 fourty-three

43 43 43 43

43 43 43 43

43

fourty-three

fourty-three

44 fourty-four

44 44 44 44

44 44 44 44

44

fourty-four

fourty-four

45 fourty-five

45 45 45 45

45 45 45 45

45

fourty-five

fourty-five

46 fourty-six

46 46 46 46

46 46 46 46

46

fourty-six

fourty-six

47 fourty-seven

47 47 47 47

47 47 47 47

47

fourty-seven

fourty-seven

48 fourty-eight

48 48 48 48

48 48 48 48

48

fourty-eight

fourty-eight

49 fourty-nine

49 49 49 49

49 49 49 49

49

fourty-nine

fourty-nine

BONUS coloring pages, have FUN!

50 fifty

50 50 50 50

50 50 50 50

50

fifty

fifty

51 fifty-one

51 51 51 51

51 51 51 51

51

fifty-one

fifty-one

52 fifty-two

52 52 52 52

52 52 52 52

52

fifty-two

fifty-two

53 fifty-three

53 53 53 53

53 53 53 53

53

fifty-three

fifty-three

54 fifty-four

54 54 54 54

54 54 54 54

54

fifty-four

fifty-four

55 fifty-five

55 55 55 55

55 55 55 55

55

fifty-five

fifty-five

56 fifty-six

56 56 56 56

56 56 56 56

56

fifty-six

fifty-six

57 fifty-seven

57 57 57 57

57 57 57 57

57

fifty-seven

fifty-seven

58 fifty-eight

58 58 58 58

58 58 58 58

58

fifty-eight

fifty-eight

59 fifty-nine

59 59 59 59

59 59 59 59

59

fifty-nine

fifty-nine

60 sixty

60 60 60 60

60 60 60 60

60

sixty

sixty

61 sixty-one

61 61 61 61

61 61 61 61

61

sixty-one

sixty-one

62 sixty-two

62 62 62 62

62 62 62 62

62

sixty-two

sixty-two

63 sixty-three

63 63 63 63

63 63 63 63

63

sixty-three

sixty-three

64 sixty-four

64 64 64 64

64 64 64 64

64

sixty-four

sixty-four

65 sixty-five

65 65 65 65

65 65 65 65

65

sixty-five

sixty-five

66 sixty-six

66 66 66 66

66 66 66 66

66

sixty-six

sixty-six

67 sixty-seven

67 67 67 67

67 67 67 67

67

sixty-seven

sixty-seven

68 sixty-eight

68 68 68 68

68 68 68 68

68

sixty-eight

sixty-eight

69 sixty-nine

69 69 69 69

69 69 69 69

69

sixty-nine

sixty-nine

70 seventy

70 70 70 70

70 70 70 70

70

seventy

seventy

71 seventy-one

71 71 71 71

71 71 71 71

71

seventy-one

seventy-one

72 seventy-two

72 72 72 72

72 72 72 72

72

seventy-two

seventy-two

73 seventy-three

73 73 73 73

73 73 73 73

73

seventy-three

seventy-three

74 seventy-four

74 74 74 74

74 74 74 74

74

seventy-four

seventy-four

75 seventy-five

75 75 75 75

75 75 75 75

75

seventy-five

seventy-five

76 seventy-six

76 76 76 76

76 76 76 76

76

seventy-six

seventy-six

77 seventy-seven

77 77 77 77

77 77 77 77

77

seventy-seven

seventy-seven

78 seventy-eight

78 78 78 78

78 78 78 78

78

seventy-eight

seventy-eight

79 seventy-nine

79 79 79 79

79 79 79 79

79

seventy-nine

seventy-nine

80 eighty

80 80 80 80

80 80 80 80

80

eighty

eighty

81 eighty-one

81 81 81 81

81 81 81 81

81

eighty-one

eighty-one

82 eighty-two

82 82 82 82

82 82 82 82

82

eighty-two

eighty-two

83 eighty-three

83 83 83 83

83 83 83 83

83

eighty-three

eighty-three

84 eighty-four

84 84 84 84

84 84 84 84

84

eighty-four

eighty-four

85 eighty-five

85 85 85 85

85 85 85 85

85

eighty-five

eighty-five

86 eighty-six

86 86 86 86

86 86 86 86

86

eighty-six

eighty-six

87 eighty-seven

87 87 87 87

87 87 87 87

87

eighty-seven

eighty-seven

88 eighty-eight

88 88 88 88

88 88 88 88

88

eighty-eight

eighty-eight

89 eighty-nine

89 89 89 89

89 89 89 89

89

eighty-nine

eighty-nine

90 ninety

90 90 90 90

90 90 90 90

90

ninety

ninety

91 ninety-one

91 91 91 91

91 91 91 91

91

ninety-one

ninety-one

94 ninety-four

94 94 94 94

94 94 94 94

94

ninety-four

ninety-four

95 ninety-five

95 95 95 95

95 95 95 95

95

ninety-five

ninety-five

100 one hundred

100 100 100

100 100 100

100

one hundred

one hundred

Don't forget your FREE goodies!

Scan this QR Code

Congratulations

on completing

T-Bot Learns: Numbers 1-100

is awarded to:

____

____

Date

Presented by:

T-Bot

Made in United States
Orlando, FL
07 November 2026

90337524R00059